Animals That Use Tools

Animals That Use Tools

Barbara Ford

Illustrated with drawings by Janet P. D'Amato and photographs

JULIAN MESSNER NEW YORK

Library of Congress Cataloging in Publication Data

Ford, Barbara.
 Animals that use tools.

 Includes index.
 SUMMARY: Discusses the ability of many different
animals including crabs, vultures, otters, elephants, and
chimpanzees to use tools and the ability of a fewer
number of species to make tools.

 1. Tool use in animals — Juvenile literature.
[1. Tool use in animals. 2. Animals — Habits and
behavior] I. D'Amato, Janet. II. Title.

QL 785.F65   591.5   78-16895
ISBN 0-671-32950-2

For my father, who loves animals
even more than I do

Acknowledgements

Many of the concepts in this book are based on the work of
Dr. Benjamin B. Beck of the Brookfield Zoo, Chicago, Ill.,
who furnished me with his scientific papers and discussed
his work with me. Dr. Alan C. Kamil of the University of
Massachusetts was also very helpful. My special thanks to
both. Thanks also to Thony B. Jones, Joan H. Fellers, Mary
Smith of the National Geographic Society, Dr. Jane Goodall
and last but not least to No. 103 for her splendid tool-using
demonstration during my visit.

Messner Books by Barbara Ford

ANIMALS THAT USE TOOLS
KATYDIDS: THE SINGING INSECTS
HOW BIRDS LEARN TO SING
CAN INVERTEBRATES LEARN?

PHOTO CREDITS

Chicago Zoological Society photo by Benjamin Beck, p. 54
Chicago Zoological Society photos by Leland LaFrance, pp. 58 to 62
Cleveland Museum of Natural History, p. 87 and frontispiece
Joan H. Fellers & Gary M. Fellers, p. 24
Alan C. Kamil & Thony Jones, pp. 28, 30, 31
Parrot Jungle - Miami, Florida, p. 15
San Diego Zoological Society, pp. 72, 77, 78, 79, 81
Baron Hugo van Lawick, ©National Geographic Society, pp. 13, 41, 70-71

Contents

Animals That Use Tools

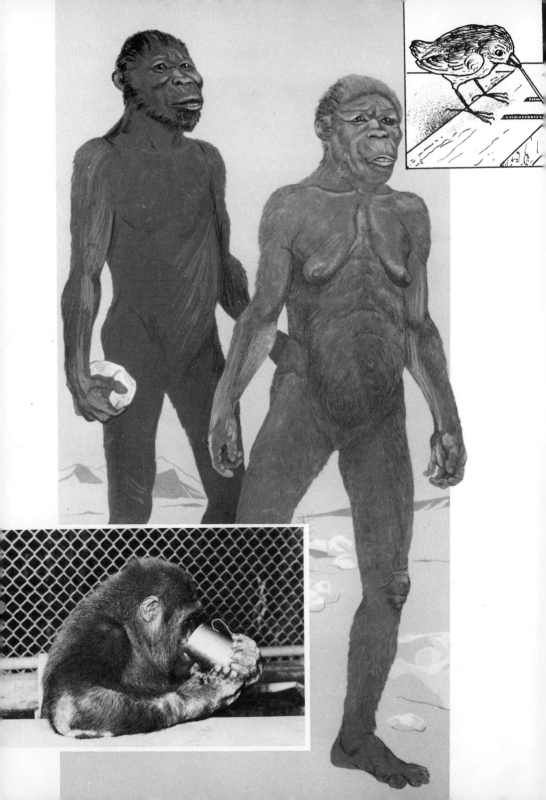

1

Animals and Tools

An ant puts some jelly on a leaf and drags the leaf home to its nest.

A bird uses a piece of paper as a broom to sweep in food.

A baboon throws a rock at an observer.

Each one of these animals is using a tool. One definition of a tool is an object outside the body that an animal picks up and uses to achieve a goal. A surprising number of animals have been seen using objects in this way. Among them are crabs, vultures, otters, elephants, and chimpanzees.

The world's number one tool user is the human. We use a spoon to eat food. We use a saw to cut wood. We use a violin to produce music. We could not exist without tools, and the evidence indicates that human beings have used tools since our early ancestors appeared on earth millions of years ago. Not only do humans use tools: we make tools. Some of the tools we make are very complex and require a pattern, or design.

Until the last century, scientists believed that the ability to make and use tools was one way in which man differed from animals. It was generally accepted that the lower animals neither used tools nor made them. Then, in the 19th century, a few scientists saw wild animals using tools. Several reports of this behavior appeared in scientific journals.

Later, the development of a new science, *ethology*, played a role in the acceptance of animal tool use. Ethology emphasizes the importance of studying the natural actions of animals in the wild. *Ethologists* — scientists who practice ethology — left their laboratories and went out into forests and fields to observe animals. Some ethologists saw and reported on animals using tools.

By the 1930s, there were a number of reliable reports published of animal tool use. A wasp uses a stone to build its nest. A crab holds another animal in its claws to defend itself. A bird picks up a stick in its beak and uses it to dig insects out of cracks. An otter breaks clam shells by pounding them on a rock held on its chest. Elephants scratch their hides with tree limbs held in their trunks.

Monkeys even throw branches and rocks at nosy scientists — striking evidence of tool use, indeed!

Today, scientists continue to see many different animals using tools. One of the most exciting examples of

animal tool use was described by Dr. Jane Goodall, a British scientist. When she was studying wild chimpanzees in Africa in 1960, she saw them use a grass stem to get termites out of a termite nest. Goodall also saw something else: tool making. The chimpanzees would occasionally take a small twig and tear off the leaves to make it fit better in a nest.

The action of tearing off leaves before using a tool is an example of simple tool making. Perhaps this is the way our distant ancestors began to make tools.

Not all scientists accept all these examples as tool use, but the evidence that at least some animals use tools is overwhelming. A few even make simple tools.

The case for animal tool use has also been strengthened by evidence from scientists who observe animals in captivity. Sometimes this is the best way to study tool use. Many monkeys, and all the apes (gorillas, orangutans, chimpanzees, gibbons) use various tools to get food when in captivity. In most cases, the animals do not have to be trained to get their food this way. Some captive birds also use tools to get food.

Recently, some captive chimpanzees have been trained to use tools to give messages to humans.

But not all the "tool-using" behavior we see in captive animals is really tool using.

Have you ever seen an animal show in which a trained animal does various tricks? I saw a show in which a rabbit banged on a little piano with its paws and a pig used a broom to sweep the floor. As soon as these animals finished these tricks, they got a reward of food. Trained animal acts like this are fun to watch, but they do not show tool use in the scientific sense.

The trained pig, for instance, does not pick up a broom and make sweeping motions because it wants to clean the floor. Pigs have no interest in clean floors. The trained pig

Pinky, a Moluccan cockatoo, is a star performer on the tightrope wire, pedaling the bicycle with its claws.

takes a broom someone gives it and moves it around because it has been trained to expect food if it does this. The food, not the possibility of a clean floor, is what makes it move the broom around. Only if the pig wanted to clean the floor would sweeping be considered tool using.

In scientific experiments, rats and other animals have been trained to press bars to get food rewards. This is much the same sort of behavior as that shown by the trained pig. The bar-pressing rat does not pick up the bar and use it to get food. Pressing a bar that is already set up is simply a sort of signal. The signal says to the researcher who designed the experiment: "Okay. I've done this strange thing you want me to do. Now where's my reward?"

But real tool-using behavior is found in many different animals, from the very simple to the very complex.

Scientists divide all the animals in the world into two large groups, the invertebrates and the vertebrates. There are many differences between them, but the basic difference is that the vertebrates have a backbone, while the invertebrates do not. Almost all invertebrates have a very small brain or no brain at all. They include the world's least intelligent animals.

Vertebrates, in general, have larger brains and larger bodies than invertebrates. Not only do vertebrates have a higher level of intelligence than invertebrates, but vertebrates often have body parts such as hands that make it easy to hold and move tools. Vertebrates are divided into five large groups: fish, amphibians (frogs and toads), reptiles, birds, and mammals. Almost all the vertebrate tool users are found in the last two groups: the birds and the mammals.

The *primates,* a special group of mammals made up of human beings, apes, and monkeys, are the world's best tool users.

Although there are many more tool users among the big, intelligent vertebrates, the invertebrates also include some good tool users. And one of these tool-using invertebrates, the snail, does not even have legs or arms! Several new invertebrate tool users have been found in recent years by American scientists.

How can an invertebrate use a tool when it does not have legs or arms? Why do more birds and mammals use tools than other vertebrates? Why are primates the best tool users? And does intelligence have anything to do with tool use? A closer look at some of the creatures that use tools will answer some of these questions — but raise others. We still have much to learn about our fellow tool users.

2

The Ant with the Shopping Bag

Dr. S. W. Williston, of Lawrence, Kansas, was collecting insects in western Kansas on a hot summer day in 1891. As he worked, he saw female sand wasps busy with some activity on the bare earth. Williston, an *entomologist*, a scientist who studies insects, stopped to watch them. Each wasp dug a hole, then flew away. First she returned with a small stone, then another, each time carrying it in her jaw. Then she carried back some larvae, an early life stage of an insect. The wasp put the larvae in the bottom of the hole, covered it with one stone, and put dirt on top.

Then, to Williston's surprise, he saw the wasp take a second stone in her jaws and use it to press down dirt over the hole. The wasp was using a tool! He saw other wasps do the same thing.

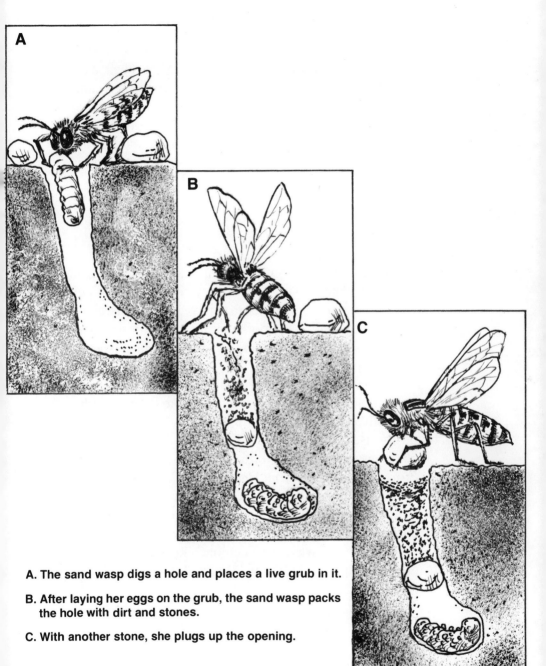

A. The sand wasp digs a hole and places a live grub in it.

B. After laying her eggs on the grub, the sand wasp packs the hole with dirt and stones.

C. With another stone, she plugs up the opening.

Today we know that the wasp had laid her eggs in the hole, before covering them up. When the young wasps came out of the eggs, they had the larvae to eat.

Were these tool-using wasps Williston saw on that day in 1891 an unusual group of super-wasps? Not at all. Although no one had noticed an invertebrate like the sand wasp using a tool before, it was not long before scientists in other areas reported the same behavior in wasps.

Once scientists were alert to the fact that at least one invertebrate uses tools, they began looking for the same behavior in other invertebrates. Early in the twentieth century, J.W. Duerden, a South African scientist, found tool using in a crab. The crab is a *crustacean*, a member of a group of invertebrate water creatures that have hard shells and a number of long legs. The tool-using crab's name is *Melia tesselata*. It lives in the waters off the Hawaiian Islands, among other places.

This crab is one of the few animals using a living tool: the sea anemone. The anemone is a flowerlike creature with a thick, stalklike base and many long arms. It cannot move around by itself, so it attaches its base to a solid surface, extends its arms, and waits for food to come by. Sometimes what comes by is not food but *Melia tesselata*. When the crab touches a sea anemone, the crab grabs it by the base, below the anemone's arms, which sting. The crab usually carries one anemone in each of the claws on its two front legs.

Held in this way, the anemone can sting the crab's enemies and capture food, which the crab promptly takes.

Exactly seventy years after Duerden had described this unusual kind of tool using, another water-living invertebrate tool user was found. It is a sea snail, a shelled creature that lives along the California coast, as well as other coastal areas. The snail has no arms or legs, but it does have a "foot," a soft, fleshy mass of tissue. When a snail moves, the foot comes out from beneath the shell and moves up and down in tiny waves, like a tractor tread. If the snail is placed upside down, the foot attaches itself firmly to some solid surface and pulls the whole animal into an upright position again.

In 1975, two Bucknell University scientists, Paul J. Weldon and Daniel L. Hoffman, showed how an animal without arms or legs can use a tool. They put the sea snails in an aquarium with a gravel bottom made up of pebbles too small to support the weight of the snail. Then they turned the animal upside down. The snail's foot came out and probed around. Not finding a solid surface, one end of the foot picked up a pebble and transferred it to the other end. The pebble moved down the foot by means of a series of small waves typical of snail movement. When the snail had a number of pebbles on its foot, it was heavy enough to shift its center of balance. Flip! Over went the snail!

If the snail happens to get hold of a long, narrow stone, it presses the stone down into the gravel almost like a shovel, and then flips itself over.

All the other invertebrate tool users that are known today are insects. Have you ever come across a number of small pits in a sandy location? They may have been dug by the

A doodlebug trap.

antlion larvae, which are known as doodlebugs. The pits are traps. When an ant comes to the edge of the pit, the doodlebug, which is buried at the bottom, hurls grains of sand in the air. This causes little landslides on the walls of the pit, the ant tumbles in, and the doodelbug eats it.

Since antlions live in many areas of the United States, you may be able to see this behavior for yourself if you locate some doodlebug pits. Dig up the doodlebugs (see illustration above) and put them in a dish of sand inside a large box. Put some ants in the dish and see what happens.

Another insect, the wormlion, a kind of fly that lives in the southwestern United States, catches food in almost exactly the same way the doodlebug does. The two insects, however, are not related in any way. As with the antlion, it is the wormlion larvae that actually uses the tool. It digs a pit in

Joan H. Fellers

sandy soil, buries itself, and tosses sand toward any small insect or worm that comes to the edge of the pit.

Perhaps the most remarkable example of tool using by an insect was discovered very recently by two American scientists.

In a 1976 issue of *Science*, a scientific journal, Joan H. Fellers and Gary M. Fellers reported that some ants use tools to carry food. These ants drop leaves, pieces of tree bark, and other material, on rotten fruit or dead insects. Then the ants leave. While they are gone, the materials soak up the food juices. When the ants return, they carry the soaked material off to their nest, much as we might carry a shopping bag full of food.

To find out what happens to the juice-filled pieces of leaves and bark, the Fellers, who are now with California State University, built a clear plastic ant nest and put ants in it. Then they put jelly on the ground near the nest. The ants found the jelly and put leaves and other material on it, returning later to take their "shopping bags" home. When the Fellers looked inside the nest, they saw many of the ants feeding from the jelly-soaked material.

The ant takes a leaf to a food source such as a piece of fruit, placing it so that it will soak up the juices. Often other ants help position the leaf.

After the leaf has absorbed the sweet juices, it is taken back to the colony so other ants can feed on it.

The ants the Fellers studied are myrmicine ants, which live in many areas of the United States. For ants, their behavior is unusual. Most ants fill a part of their body called the *crop* with food wherever they find it, and carry this food back to the nest. Why don't the myrmicines do the same thing?

Near the myrmicines live a number of big, aggressive ants which have been seen chasing the small myrmicines away from food. The Fellers think the smaller ants developed their unusual method of carrying food so they can get food quickly, before they are attacked. On several occasions, the myrmicines were seen putting a piece of material down on food the aggressive ants were eating, then scurrying away. Then the myrmicines returned and took the soaked material away when the bigger ants weren't looking!

As we'll see throughout this book, many tool-using animals seem to have hit upon tools as a way of competing successfully with other animals.

The Jays that Sweep
with a Broom

A captive female blue jay perches on a bar in her cage. Just outside the front of the cage are a number of pieces of food, but the jay cannot reach them. Thony B. Jones, a graduate student at the University of Massachusetts where these experiments are being conducted, puts some small strips of white paper in the top of the jay's cage. The strips are about two and one-half inches long and one-quarter inch wide. He does the same thing with the cage of a male jay a few feet away from the female.

As soon as Jones moves away, the female grabs one of the strips with her bill, hops down to her water dish, and dunks the strip in the dish. Then she walks over to the front of the cage and pokes the wet strip of paper through the bars.

3

Dr. Alan C. Kamil, left, Thony Jones, right, and Number 103.

She moves the paper back and forth, sweeping at the food in front of her cage until she moves it within reach of her bill. She eats the food, then moves to another part of the cage and repeats the performance.

Watching the tool-using bird are Dr. Alan C. Kamil, a psychologist who co-designed the experiment, and a visitor. "Number 103 hasn't done this in four years," says Kamil. "But she remembered."

While Number 103 is using its paper strip to get food, the other jay, which has also grabbed a paper strip, is crumpling it with his foot and beak. Sometimes he pushes the paper out of the front of the cage, but he doesn't sweep it back and forth to move the food toward the cage. After about a half-hour, though, he manages to sweep a piece of food within reach. Soon he gets more food, but he never uses his strips as efficiently as his neighbor.

"Both these birds had experience doing this four years ago," says Jones. "But Number 103 did it about 100 times while the second bird, Number 114, did it only about 30 times. Maybe that's why 103 remembers better."

Jones was the first to notice that blue jays are able to use a paper tool. In the early 1970s, Kamil was doing experiments with blue jays that did not involve tool use. He captured wild blue jays at 17 to 21 days of age from nests around Amherst, where the University is located, and reared the birds in his laboratory. (It is against the law to capture and rear native wild birds like the blue jay, so Kamil first obtained a Federal permit to enable him to do this.) Jones helped care for the birds and conduct experiments.

One day, when Jones was in the room where the blue jays are housed, he saw one of them tear a strip of newspaper from a large piece lining the bottom of the cage. The bird twisted the strip with its beak and feet. Then it walked to the front of the cage, pushed the strip out, and swept it back and forth. Some pieces of food that had been dropped accidentally lay outside the cage, and the paper swept some within reach of the bird, who quickly ate it.

A

B

C

Preparing to sweep up food . . .

D

E

F

. . . . successfully!

31

Jones could hardly believe his eyes. "I guess I'll never see that again," he told himself. He went out and came back a half-hour later. The bird was still sweeping up bits of food with the strip of paper!

"Thony came running down the hall into my office and said: "I think we've got a tool user," Kamil remembers.

Actually, they not only had a tool user but a tool maker. The bird had made the strip she needed from the larger piece of paper in the cage. Kamil and Jones made a film of the bird's performance and wrote a scientific paper that was published in the journal *Science*. The paper describes how the bird made paper strips and used them and also used objects like feathers, a piece of straw, a paper clip and a plastic bag tie to sweep food within reach.

The paper also gives the results of an experiment Kamil and Jones carried out with eight other blue jays in their colony. Five of the eight used paper strips to obtain food without any training.

Later on, Kamil and Jones carried out another experiment with 20 jays, ten of which had an opportunity to watch jays who were tool users. The other ten birds did not see the tool users. But the results in both groups were similar. In each group about the same number of birds learned to use paper strips to get food. In all, eleven birds of the 20 showed tool-using behavior.

Kamil believes that one of these eleven birds may have shown a special kind of learning called "insight." People or animals show insight when they suddenly see a

connection between two things, such as a tool and food. This happens without their having to perform some action over and over to learn it. When the bird that Kamil believes showed insight first saw the paper strip, it looked at it, then looked at the food. Then it looked back and forth from the paper strip to the food a number of times. During this time, it did not try to use the strip. But suddenly, it grabbed the strip, walked to the front of the cage, and used the strip to sweep a piece of food within reach.

Was this insight? Kamil doesn't know for sure, but he says that it certainly looks like insight.

The rest of the birds in the study seemed to learn how to use the paper strip by a type of learning scientists call "trial and error learning." They played with the strip for a long time, crumpling it under their feet and pecking at it with their bill. Blue jays often play with objects in this way. Sometimes they pushed the strip out of the front of the cage, then pulled it back in. Sometimes, by accident, a piece of food was pushed this way to within reach of the bird's beak. After this happened a number of times, some birds seemed to realize what the strip could do. They then began to repeat the sweeping movements deliberately.

Many scientists believe that most, if not all, tool using is learned in this same way by animals. In Chapter 6 there is an account of a monkey that showed much the same kind of behavior as the jays.

If you have a parrot or parakeet, Kamil suggests you try an experiment to see if it can use a paper tool to get food. Parrots and parakeets are not related to jays, but they move small objects around with beak and feet the way jays do. Put some of the bird's favorite food outside its cage where it cannot reach it except with a tool. Then cut some strips of paper and put them in the bars of the cage.

Don't give up after you try this once or twice. It may take the bird many trials to figure out how to use a tool.

Do jays use tools in the wild? No scientist has seen such a thing, but it may happen. A California reader of a popular magazine in which an article appeared about the captive jays reported that he had seen a wild jay using a tool. The bird pried a nut out of a crack with a large twig. "Animals use tools much more than we used to think they did," says Alan Kamil. He thinks that if more people watched birds in the wild to see if they were using tools, they would probably see this behavior from time to time.

If you have any wild blue jays around your home, you may see them using a tool yourself. If you do, write Dr. Kamil at the Department of Psychology, University of Massachusetts, Amherst, Mass. 01003. Describe the bird's tool-using behavior in detail, and include photographs of the behavior, if possible.

4

Little Bird with A Big Stick

The jay is just one of many birds that use tools. A vulture and a buzzard drop stones on eggs to open them. A finch and a nuthatch use a stick or piece of bark to push live food out of cracks. The bowerbird of Australia paints its nest using a brush made out of bark. A heron has been seen using bait to catch fish, both in the wild and in a Miami zoo. At least a dozen different kinds of captive and wild birds have learned to pull up food by means of a string.

And all these birds use tools, not just occasionally, but as a part of their regular activities.

Probably the most interesting bird tool user in the world is one that most of us will never see. It is the Galapagos woodpecker finch, a small bird that lives on the Galapagos Islands 600 miles from the western coast of South America.

These were the islands that Charles Darwin, the British scientist, talked about in his famous book, *The Origin of the Species*. Because these islands have been separated from the mainland for such a long time, animals developed on the islands that are not seen anywhere else in the world. The Galapagos woodpecker finch is one of these unusual creatures.

A tree-living finch, this bird climbs trees like a woodpecker, looking for insects. We are familiar with woodpeckers in the United States that push their long tongues into cracks in trees to get live food out. The Galapagos woodpecker finch does not have a long tongue, so it uses a twig to do the same job. When it finds a worm or insect in a crack, it

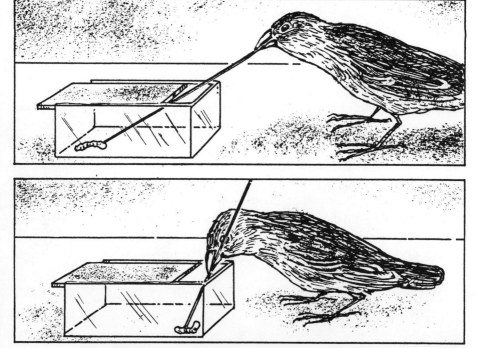

For laboratory observation, a mealworm was placed in a transparent box, the top opened slightly. The finch found the opening and inserted the tool to reach the worm. When the worm was far back, the finch held the tool at the end. When the worm was close to the opening, the bird held the tool in the middle.

breaks off a twig from a branch with its bill. Then it moves the twig back and forth in the crack, forcing the food out.

Most of the time, the woodpecker finch picks a twig the right size for the crack. But an American scientist, Robert Bowman, saw one of these Galapagos birds break off part of a twig that was too long. This is an example of simple tool making.

To study the behavior of the Galapagos woodpecker finch more closely, Bowman and another American scientist, George C. Millikan, brought some of these birds back to their laboratory. There the birds used tools ranging from a toothpick to a piece of uncooked spaghetti to find food. If a tool was very long, the bird stretched its neck out as far as it could and held the tool at the end. If the tool was very short, the bird bent over and held the tool in the middle. On one occasion, a woodpecker finch threatened another bird with a metal stick. This kind of threat with a tool had never been seen in birds in the wild.

In this laboratory experiment, a mealworm was placed in a slot. The finch inserted the tool toward the back of the slot and pulled it forward, dislodging the worm.

As far as we know, the Galapagos finch is the only bird to ue a stick to obtain food, but a little bird that lives in parts of the United States does something similar. The brown-headed nuthatch has been seen using a piece of pine bark to pry up other pieces of bark on trees. Underneath the bark are insects, which the nuthatch eats. A bird four to five inches long with a brown cap on its head, the brown-headed nuthatch lives in pine woods from Florida and the Gulf of Mexico north to Delaware in the east and Missouri in the west. If they live in your area, see if you can spot them using a bark tool.

At least two different birds have been seen dropping or throwing stones on food. Jane Goodall, the British scientist, saw Egyptian vultures, crow-sized African birds, dropping stones on ostrich eggs before eating them. Ostrich eggs are big and have a tough shell. The birds picked up a stone

A nuthatch prying up bark to get at grubs.

with their bill, raised their heads high with the bill pointing upward and threw the stone down hard on the egg. It took four to twelve throws to break the egg. Photographs of this behavior appear in *National Geographic Magazine* for May, 1968.

The other stone-dropping bird, the black-breasted buzzard of Australia, flies above the eggs and drops stones down on them. It breaks at least three different kinds of eggs this way.

Sometimes the behavior of a bird tool user is so much like man's that it is hard to believe unless you see it. A green

An Egyptian vulture getting ready to crack an ostrich egg.

A. The heron finds bait . . .

B. . . . and takes it down to the water.

heron in the Miami Seaquarium, a zoo, fishes with bait. The heron, a crow-sized bird with a greenish-blue head, takes some pieces of food and scatters them on the water. When fish come to snap at the food, the heron grabs the fish and eats them. The heron's mother and brother fish the same way, but not as often. Photographs of the fishing herons have appeared in *National Geographic Magazine*.

Actually, the Miami Seaquarium herons are not the only herons known to fish with bait. In a 1957 scientific journal, scientist H.B. Lovell describes a wild green heron using bait to catch fish.

C. It drops the bait in the water . . .

D. . . . then waits for a fish to come by.

As a fish goes for the bait, the heron pounces.

F. Dinner!

Two kinds of bird behavior look very much like tool using but they are not accepted by all scientists as tool use. One is string pulling, the other shell dropping. Many captive and wild birds learn, without any training, to pull up food dangling from the end of a string. All the birds perform this feat in much the same way. First they grasp the string with their beak and pull, then they put the loose string under their foot. This keeps on until the food is close enough to grab with the beak.

One persistent bird pulled up a string 25 inches long this way!

Two European species, the titmouse and the goldfinch, are particularly good at string pulling. In Europe, people used to keep goldfinches in a cage in which they could obtain food and a small container of water by pulling up a string for each. Altogether, string pulling has been seen in a dozen different species, including the parrot and budgerigar or "budgie." If you have one of these birds, you may be able to see this behavior by tying food to the end of a string and hanging it from a perch inside the cage. Or try the same arrangement at an outdoor feeder to test wild birds.

String pulling is not accepted as tool use by all scientists because the experimenter, not the bird, ties the string to the food. In other words, the bird does not make the physical contact between the tool (string) and the food.

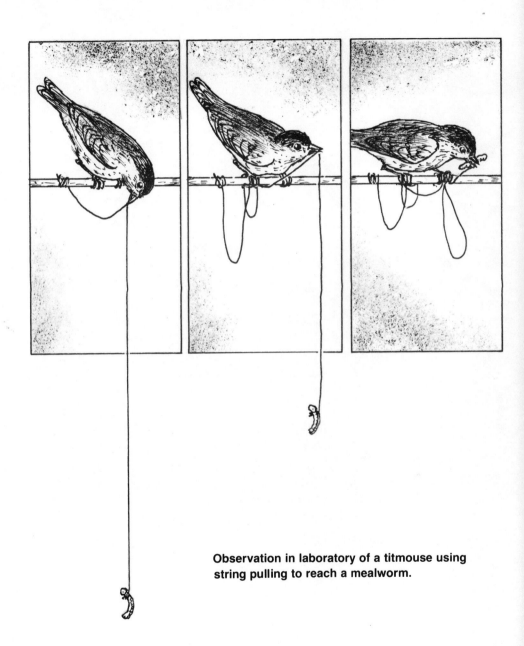

Observation in laboratory of a titmouse using string pulling to reach a mealworm.

Shell dropping is not accepted as tool use by some scientists for a different reason. In shell dropping, the bird picks up a shell, flies above a hard surface such as a parking lot, and lets the shell drop. The shell breaks, allowing the bird to eat the soft-bodied animal inside. But since the bird does not actually pick up the hard surface, shell dropping does not quite fit the definition of tool use given in the first chapter.

Many different birds drop shells, including crows, bald eagles and sea gulls. Gulls are particularly noted for this behavior. According to an American scientific journal, a policeman named Abe Locke was walking along a boardwalk in Atlantic City, N.J., in 1924, when he was struck on the head by a clam dropped by a sea gull. Locke was knocked unconscious and had to be carried to a drugstore to be revived.

All of the birds described so far use tools to get food. One bird uses a tool for a more artistic reason: to paint its nest. The male satin bowerbird of Australia, a handsome bird with shiny blue-black feathers, takes a wad of bark in its beak and uses it to paint its bower, or nest, with a mixture of charcoal and the bird's saliva. Bowerbird nests are large, colorful structures, decorated with flowers and bits of shiny objects the birds can find, such as coins and keys. The males build these fancy nests to attract females.

Bowerbird decorating its nest.

Why do so many birds use tools? Many scientists now believe birds are more intelligent than we used to think they were, but intelligence probably has less to do with tool use than with other abilities in animals. The ability to hold and move objects that can serve as tools probably plays a bigger role in tool use than intelligence. So does the time an animal spends using tool-like objects. In species of birds which use tools, individual birds are often seen playing with objects that might serve as tools.

Jays, for instance, often play with objects. This kind of behavior may set the stage for tool use in jays once they see the connection between using a tool and gaining a desired object.

The conditions under which animals live may also help determine which of them use tools. The Galapagos woodpecker finch lives in a climate where insects crawl into cracks during the day. If the finch could not pry them out with a tool, it might not be able to live on these hot, dry islands. The Egyptian vulture is the smallest vulture in its area and bigger vultures often drive it away from food. By breaking eggs with stones, this small vulture makes sure it can grab some food before other birds get there.

5

The Living Water Pistol

One of the world's most unusual animal tool users is a fish. The archer fish, a species native to waters from India to Australia, swims just below the surface, near the banks of swamps and streams. When it sees an insect on vegetation above the water, it shoots out a stream of water from its mouth. If the fish's aim is good, the insect drops into the water, where the fish eats it. The archer fish can hit an insect up to four feet away.

To shoot out its stream of water, the archer fish suddenly squeezes its gill covers, the protective fold of skin over the gills. This movement forces water through a tube formed by a groove in the roof of the fish's mouth and its tongue, and so out the mouth. It works much like a water

pistol. In fact, European scientist K.H. Luling, who has studied the archer fish, calls it "the living water pistol."

The archer fish is a popular aquarium fish in the United States. If you have one, try dangling a mealworm on a thread above the water to see if it will "shoot" the food. But watch out! The archer fish will sometimes shoot any object on or above the water, so you may get a stream of water in the face if you simply bend over the aquarium.

The archer fish is the only fish tool user and the only vertebrate tool user that is not a bird or a *mammal*. Mammals are hairy vertebrates that feed their young with milk. Most of the mammal tool users are primates. There are only three known exceptions: the California sea otter, the elephant, and the polar bear. And only *one* of these three mammals — the sea otter — uses tools on a regular basis.

There are reports, now and then, of other mammals using tools, but almost all these reports involve only one or two unusual individuals. One horse, for instance, has been photographed scratching his back with a stick. But in all the thousands of years that man and horse have been associated, only a few unusual individual horses have been reported to use tools. Tool using, in other words, is not typical of horses.

On the other hand, many members of the three mammal groups named above have been seen using tools. Tool using is typical behavior for them. In the case of the elephant, which has been associated with man as long as the horse, a number of different elephants, both in the wild and in captivity, have been seen using a stick to scratch themselves. In

Accurate aim is essential—this archer fish gets only one chance at a target.

A big back—a big back scratcher!

their book *Among the Elephants*, Dr. Iain Douglas-Hamilton and his wife, Oria, describe a wild African elephant they saw scratching its back with a stick held in its trunk.

Both captive and wild elephants have also been seen using their trunks to throw objects and wave branches over their heads in a threatening way.

The ability of elephants to pick up objects and move them around with their trunk, together with their great strength, has been of help to man. In some Asian countries where wild elephants are found, elephants are trained to move heavy objects such as logs with their trunks. Each elephant is guided in the task by a human master.

The evidence for tool use in the polar bear isn't quite as good as for the elephant, partly because there are far fewer inhabitants in the Arctic, polar bear territory. Eskimos in many areas of the Arctic have seen polar bears stand on their hind legs and throw chunks of ice at walruses. No scientist has seen this behavior in the wild as yet. But in captivity, polar bears toss heavy objects, including ice, around their enclosures. Scientists have also seen wild polar bears stand on their hind legs and handle objects with their front paws.

The best new evidence for polar bear tool use comes from Dr. Benjamin Beck of the Brookfield Zoo in Chicago, Illinois. In 1976, he told a scientific conference that a female polar bear in the Brookfield Zoo often picks up an aluminum beer keg, holds it between its front leg and the side of its head, and walks up a flight of stairs. Then, standing on its hind legs, it throws the keg down on the ground. If a walrus were there, the bear would probably hit it. Beck showed movies of this behavior to the scientists at the conference.

The Brookfield bear's behavior is so much like what the Eskimos describe, that Beck believes the Eskimo's description of wild polar bears throwing chunks of ice at walruses is accurate.

The best tool user among the non-primate mammals is not the ice-tossing polar bear or the stick-holding elephant, but the little California sea otter. In 1963, Dr. K.R.L. Hall and Dr. George B. Schaller spent a week observing otters in the Pacific Ocean off Point Lobos State Park, California. To see the otters, they used binoculars and a spotting scope, a telescope that enlarges small distant objects more than binoculars do.

The two scientists were able to see the otters floating on their backs, banging shells on stones they held on their chests. After a number of blows on the stone, the shells cracked and the otters ate the flesh inside. To open a small mussel shell took the sea otter an average of 35 blows.

Sometimes an otter kept the same stone in its armpit while it dived for other shells. This is one of the few cases where an animal is known to obtain a tool and keep it for a time before it uses it.

Hall and Schaller found that otters get most of their food by banging shells on the stone tool. They think there are several reasons why otters are such good tool users. One is that the young stay with their mothers for a long time, watching her as she eats and carries out other activities. In this way, the young may learn to imitate the mother's shell-opening behavior. Another possible reason for the otter's tool use is that otters often pick up objects and move them around in a playful way. Behavior like this seems to lead to tool using in some animals.

Beware—tool-using polar bear!

But as far as we know, other species of otters do not use tools. Why not? Dr. John Alcock of the University of Washington may have an answer to this question. He points out that the sea otter is the only otter that lives in the ocean. There are many mammals in the ocean, all of them bigger than the otter. By using tools to open shells, Alcock believes, the otter can compete better with these bigger mammals.

Caution:
Rock-Throwing Monkeys

A big male baboon with a mane of long, gray hair sat in a cage at the Brookfield Zoo in Chicago. Sitting beside him was a smaller female baboon. Baboons are members of the monkey family. The two animals were picking small bits of dead skin, insects, and other objects out of each other's fur, an activity called *grooming*. A pan of fresh vegetables and fruits was outside the cage, but it was too far away for them to reach.

Suddenly the female baboon stopped grooming the male, and walked through a small door into the next cage. The door was so small that the big male could not get through it. The female picked up a long steel pole with a hook on one end

57

Dr. Benjamin Beck sets up feeding experiment at Brookfield Zoo.

that was lying on the floor. She returned to the male and put the pole down beside him. He picked it up immediately, walked to the front of the cage, and used the pole to pull in the pan of food. Then both baboons ate.

Dr. Benjamin Beck, the scientist whose work with polar bears at the Brookfield Zoo was described in the last chapter, was watching the experiment. He turned to his assistant, a student. "What did you see?" he asked.

"The female got up and went into the other cage and brought the tool back."

"That's what I thought I saw," said Beck.

When Beck described this scene later, he said he was so excited that "the hair stood up on the back of my neck." He had set up the experiment five days earlier, hoping to see cooperation between monkeys in using a tool. The male baboon in Beck's cooperation experiment had learned to use a tool to get food in an earlier experiment. But no one had ever seen monkeys cooperating in using a tool, and Beck didn't really expect to. Now, after almost a week, the two baboons were cooperating!

Not only did the monkeys cooperate in using a tool, but the female monkey seems to have shown insight. Insight,

as we saw in Chapter 3, is a kind of learning in which an animal suddenly seems to realize how to solve a problem. It usually happens when an animal is doing something unrelated to the problem. This baboon was grooming when she suddenly got up and brought back the tool. She had never used the tool herself, but she had watched the male use the tool. Did she suddenly get hungry and realize she could have food by giving the male the means of getting it?

"I think this is as clear evidence of insight as you can get," says Beck.

Benjamin Beck has been observing tool use in captive monkeys at Brookfield since 1970. He has found some interesting differences in the way monkeys use tools. The monkeys in his experiment that cooperated in using a tool are of a species called the Hamadryas or sacred baboon, from Africa. In the first experiment he set up at Brookfield, only one of the

Hamadryas baboons in a group of six learned how to use the long steel pole to get the box of fruit and vegetables. When this animal was taken away, none of the other baboons could get the food.

Beck set up an experiment with another kind of African baboon, the Guinea baboon. One baboon learned how to get the food after many hours. When it was taken away, it took much longer for the other baboons to learn how to get the food than it had taken for the first baboon to learn. Beck believes these two experiments show that baboons do *not* learn to do something by watching other baboons or other animals and then imitating what they do.

However, Beck found that another monkey, the macaque (mah-kawk′) monkey, does learn how to use a tool by watching other macaques. In an experiment, one macaque threw the hooked tool to get food in exactly the same way as another macaque did.

Many monkeys, including baboons and macaques, have been seen using tools in the wild. Jane Goodall saw olive baboons in Africa using stones and other objects to wipe their faces. A group of researchers from the University of California was stoned by another baboon, the chacma baboon, in Africa. The baboons took the stones from the walls of a canyon and dropped them down on the researchers with an underhand motion, making everyone dodge to avoid getting hit.

Another monkey that has often been seen using tools is the capuchin, the "organ grinder's monkey." In the past, these little monkeys were dressed up in a coat and hat and given a cup to hold. Then, while a man turned the handle of a portable musical instrument called the hand organ, the monkey would use the cup to collect money from the people around. We seldom see capuchins doing this today. But captive capuchins have been observed "painting" with brushes, using sticks and piling up boxes to get food, and even using a tool to pull in another tool which they use in turn to get food. They've also been known to throw objects at the heads of people they disliked!

An old engraving of an organ grinder and his monkey.

A capuchin monkey in the wild.

Wild capuchins in South America drop branches on people, throw objects, and use sticks to hit things.

Other monkeys that have been seen using tools, either in the wild or in captivity, are the howler, the red spider, the colubus (kol-oh-bus), the saki, the squirrel, the mangabey, the guenon (guh-non) and the patas.

Why are monkeys such good tool users? Monkeys are primates, the group of animals that includes the world's best tool users. Benjamin Beck and many other scientists now think that the most important reason why primates use tools so well is their hands. The hand of most monkeys and all the apes has four long fingers and a thumb, like man's hand. The fingers bend easily, so that the hand can hold objects. The monkey-ape hand is not as well-suited to using tools as man's hand, but it can pick up objects and move them around easily. Most primates do, in fact, pick up and handle objects with their hands, both in the wild and in captivity.

Another advantage primates have in using tools is their *posture*, their usual way of sitting, standing and walking. Have you ever seen a monkey in a zoo sitting upright while it ate, or played with an object in its hands? This upright posture is the ordinary way of sitting for monkeys, and it frees the monkey's hands to hold and move objects. Most monkeys and all the apes can also stand or walk on their back legs for a short distance. Some have been seen using tools while standing or walking upright.

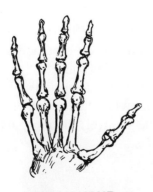

HUMAN HAND

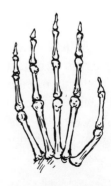

MONKEY HAND

7

The Best Animal
Tool User

One October day in 1960, Jane Goodall saw a chimpanzee use a tool in the Gombe Stream nature preserve in Africa. The animal, which she had named "David Greybeard," took a long grass stem and poked it into a termite nest. When the chimpanzee drew out the stem, it was covered with termites, which David Greybeard promptly ate. This went on for about an hour, as Goodall watched with great excitement.

There was good reason for her excitement. Wild chimpanzees in Africa had been seen using tools on a few occasions, but as far as scientists knew such behavior was very rare.

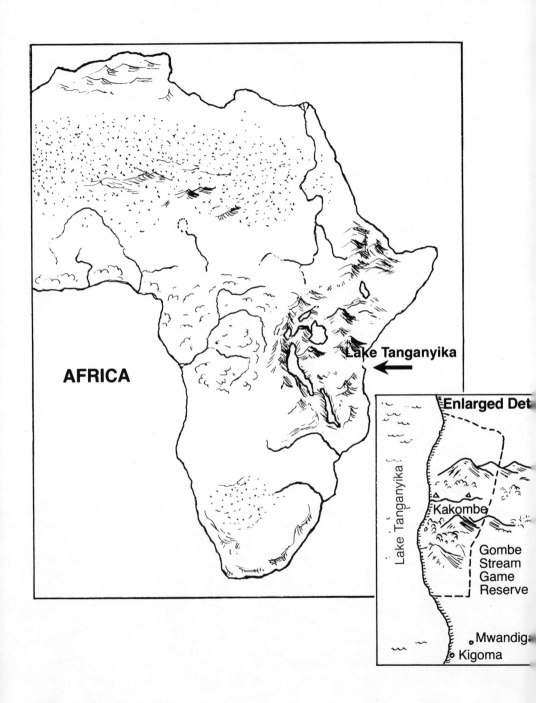

AFRICA

Lake Tanganyika

Enlarged Det

Lake Tanganyika

Kakombe

Gombe
Stream
Game
Reserve

Mwandig

Kigoma

Goodall was at Gombe Stream to study chimpanzees in the wild, a subject on which little work had been done at that time. Gombe Stream is a particularly good place for such a study. The area, which is on the shores of Lake Tanganyika in the country of Tanzania, is cut off from cities, roads and other aspects of civilization. A large group of chimpanzees lives at Gombe Stream, along with many other wild creatures.

Goodall followed the chimpanzees around on foot, using binoculars to see what they were doing. She spent all day, every day, doing this. Every night, she wrote up an account of what she had seen. She first saw David Greybeard using a tool after several months at Gombe Stream. Would the behavior be repeated? Eight days after she had first seen tool use, she had the answer. David Greybeard arrived at the termite nest with another chimpanzee. The pair worked there for an hour, using various tools to get termites.

As Goodall watched, she realized that something even more exciting than tool use was going on: tool making! On several occasions, the two chimpanzees picked up small twigs and pulled off the leaves before putting the twigs into the nest. They also bit the ends of their tools off when they became bent.

After these first observations, Jane Goodall saw chimpanzees at Gombe Stream use tools many times. The stick was the chimpanzees' favorite tool. Besides using sticks to get out termites, they used them to open boxes of food in Goodall's camp, to touch other animals and objects, to enlarge a bee's nest, and to throw at and hit other animals. One chimpanzee used a small stick as a toothpick.

Digging for food.

Sometimes the chimpanzees picked up a stick in one place and carried it to another. Few animals think ahead in this way by obtaining a tool and keeping it for a while before using it.

Goodall saw other chimpanzee tool use at Gombe Stream. They threw objects like rocks at people and animals. On a later trip to Gombe Stream, one of Goodall's student assistants was slightly injured by a two-pound rock thrown by a chimpanzee. Luckily, the rock struck the student in the leg and not the head.

Washing up. This is a captive chimpanzee.

The animals also used a sponge made of leaves to soak up water, and leaves and other materials to wipe sticky substances off their bodies. The chimpanzees made the sponge by crumbling up leaves and chewing them. This made the sponge hold more water. Most of the other kinds of tool making she saw involved ways of getting food. Besides pulling the leaves off twigs for this purpose, the chimpanzees pulled strips from blades of grass and pieces of fiber from bark. They also bit the ends of sticks to make a tool to pry open boxes.

Most of the behavior Jane Goodall saw in the Gombe Stream chimpanzees has also been seen in captive chimpanzees. The most interesting study of captive chimpanzees was made by a young German scientist, Wolfgang Kohler, during World War I. Kohler was in the Canary Islands, near the coast of Africa, when the war broke out in 1914. Unable to get back to Germany, he decided to stay on in the Canaries and study a colony of chimpanzees kept in a large enclosure.

Kohler set up a number of problems for the animals to solve, most of which involved getting food by means of tools. For several problems, food was hung from the ceiling, far above the chimpanzees' heads. Then boxes or long poles

A typical challenge set up by Kohler. How will the chimp get the basket of fruit hung out of its reach?

Problem solved!

were placed in the enclosure with the chimpanzees. In other problems, the food was placed outside the enclosure and sticks were placed inside. Sometimes the stick was in two parts. Only by fitting the parts together could the chimpanzees make a long enough stick to reach the food.

Some of the chimpanzees solved all these problems without any training from Kohler.

The animals piled up boxes and climbed poles to reach food on the ceiling and used a stick to sweep in food outside the enclosure. They also made a tool by fitting two sticks together to get the food that was out of the reach of one stick. Kohler also saw other kinds of tool making. The chimpanzees tore leaves off branches and pulled slats off boxes before using them to reach food. They took sand and rocks out of boxes before moving them to climb on to reach the food.

Another solution to the problem of getting food.

Other scientists have set up the same kind of experiments Kohler did. One scientist discovered that the age of a chimpanzee has much to do with its ability to use tools. Young adult chimpanzees do best on the kind of problems Kohler set up. Another scientist found that giving chimpanzees a chance to play with sticks and boxes for a few days makes them much better at learning how to use them for tools.

Today chimpanzees are playing the star role in other experiments which involve tools.

In some of these experiments, chimpanzees have been trained to use tools to "talk" with scientists. In a study in Georgia, chimpanzees use a computer with a special typewriter-like keyboard on which they can give and receive messages. The goal of this experiment is to communicate with chimpanzees, not to show they can use tools. But it is the chimpanzee's tool-using ability that makes this experiment possible.

Maggie, from Sumatra, brushing her teeth like the civilized orangutan she is!

All the other apes use tools, too, although not as well or as often as chimpanzees. Orangutans do many of the same things chimpanzees do with tools. Wild and captive orangutans drop and throw objects at people and animals, use sticks for various purposes, and use leaves to wipe their faces. They crumple the leaves before using them, so they also make tools. In captivity, orangutans climb poles to get food and pry open boxes with various tools.

Lowland gorilla "Bata" enjoys a cup of water.

Bata spoons it in happily.

Gorillas and gibbons use tools much less often; but both have been seen using tools in the wild and in captivity. In fact, gorillas may have a tool-using capacity almost as good

Lowland gorilla "Albert": If one hand doesn't do it, use a foot!

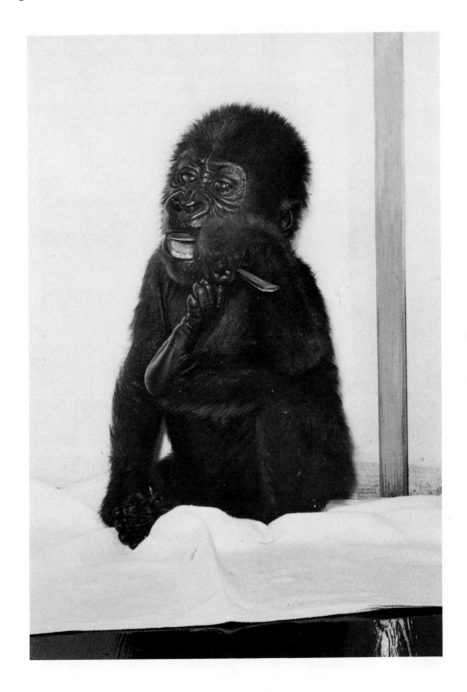

as that of the chimpanzee. A young captive gorilla in California is being trained to operate a simple computer like the one being used by the test chimpanzees in Georgia. He has learned to send and receive messages.

Why is the chimpanzee the champion tool user of the animal world?

Like other primates, chimpanzees have hands that make it easy to use tools, and they often play with objects that could be used as tools. But Benjamin Beck believes that it is the chimpanzee's ability to imitate actions that makes it such a fine tool user. The chimpanzee mimics better than any other animal except humans. We use the chimpanzee's imitative ability to train them to do tricks in circuses and shows, but chimpanzees imitate well on their own, too. People who have raised chimpanzees in their homes report they pick up many actions without being trained.

Jane Goodall tells us that young wild chimpanzees often watch their mothers closely as the mother uses a tool. Sometimes the young animal repeats her actions right afterward.

Benjamin Beck thinks this is how wild chimpanzees develop their wide range of tool-using behavior. One animal learns to use a tool by means of trial and error, which is the way most animals learn to use tools. The other chimpanzees in the group learn to use the same tool by watching the first animal. The chimpanzees pass this knowledge on to their children. Within a few generations, whole groups of chimpanzees in a certain area are using tools in the same way.

8

Early Humans, the Best Tool Makers

The first tools made and used by early humans were probably sticks like those used by the Gombe Stream chimpanzees. Stick tools, however, did not last long enough to come down to us, so the first tools we have from early man are made of stone. The oldest found so far are about two and one-half million years old and come from the shores of Lake Rudolf in Kenya, East Africa.

Similar stone tools about two million years old come from another part of East Africa, Olduvai Gorge in Tanzania.

These old stone tools are very simple. The most common one is a rounded lump about the size of a tennis ball. To make it, someone must have held a suitable rock in one hand and struck off pieces with another rock. The sharp-edged rock left after the pieces fell off could be used to cut, chop and scrape animal food. The sharp-edged pieces struck off the rock could also be used as tools.

What kind of humans made these tools?

For a long time, no one really knew. Most of the earliest stone tools have been discovered by members of one family, the Leakeys. Louis B. Leakey and his wife, Mary, found numerous stone tools in Olduvai Gorge over a period of 30 years before they found the bones of human-like creatures. Beginning in 1959, however, they found the bones of creatures that looked something like humans. The bones were near the tools and were as old as the tools or older. In all, the bones of three different man-like creatures have been found at Olduvai by the Leakeys.

The Leakeys' son, Richard, and other scientists found the remains of the same human-like creatures in other parts of Africa.

Richard Leakey has written a book, *Origins,* about early man. In his book, he argues that all three of the man-like creatures found in Africa lived at the same time. The first probably appeared about five million years ago. Eventually, two of these creatures, which are called Australopithecines, (Aus-tray-low-pith-eh-seens), died out. The third creature survived. It had more human-like characteristics than those that died out, and became our ancestor.

This creature has been given the name *Homo,* which is the first part of the Latin name for our own species, *Homo sapiens.*

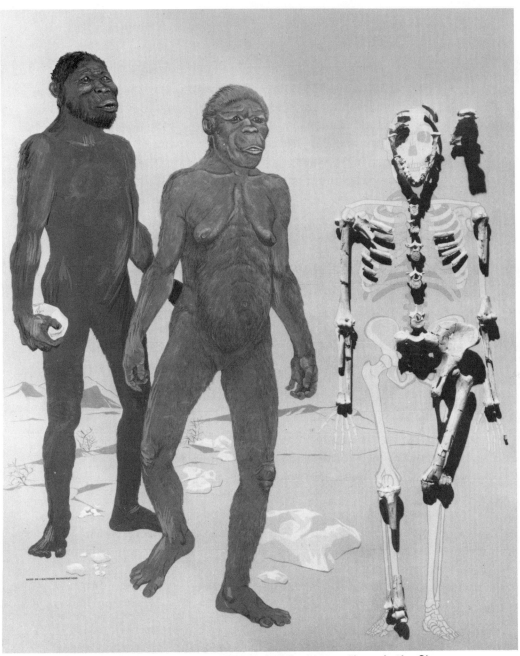

The skeleton is Lucy, named after the Beatles' song, "Lucy in the Sky With Diamonds." She is a 3,000,000-year-old Hominoid female about 4 feet tall. She seems to be related to the illustrated South African *Australopithecines* on the left.

Homo had a bigger brain than the Australopithecines but all three of these man-like creatures must have looked very much alike. All were quite short — only four to five feet tall, and all walked upright. Compared to us, they all had large jaws and low foreheads. There was one important difference in the way they looked, however: their hands. The Australopithecines had a hand with a shorter thumb than that of *Homo*. *Homo* probably had a thumb almost as long as ours.

The length of the thumb and the way it moves play an important role in tool making. To make complex tools, you need a certain kind of thumb, the "opposable thumb." Bring your thumb together with the tip of your first finger. Now bring your thumb together with the tips of your other fingers, one by one. Easy? Of course. For humans, this movement, called *opposition, is* easy. No other primate, however, can perform this movement as well as we can because the thumbs of other primates are too short, or do not move as easily as ours.

With our opposable thumb, we can grip objects in such a way that we can make precise movements. Such movements are required for shaping complex tools. This kind of grip is called the "precision grip."

Primates that do not have an opposable thumb depend on the "power grip" instead of the precision grip. With the power grip, a primate can hold an object firmly in one hand to perform movements such as hitting another object. A few non-human primates also use a kind of precision grip that

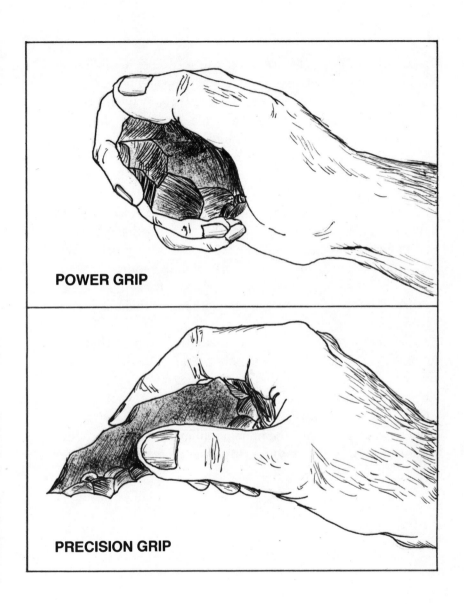

POWER GRIP

PRECISION GRIP

does not allow for the very precise movements of human beings. Humans can use *both* the precision and power grips. The earliest stone tools were made by means of the power grip, so they could have been made by any of the three man-like creatures that lived millions of years ago in Africa.

More complex stone tools have also been found at Olduvai Gorge. These tools are perhaps a million years younger than the oldest stone tools. Since these tools required more work, they were probably shaped by *Homo*, not by the Australopithecines. *Homo* probably did not have a completely opposable thumb, but it was close enough to our own to allow our ancestor to make fairly complex tools.

The earliest tools found in East Africa show, however, that you do not need a human hand to make a simple stone tool.

Modern human beings are still the world's best tool makers and users. We are the only creatures that make complex tools. We are the only creatures that make tools in a set way, and to a regular pattern. And we are the only animal that uses sources of energy outside the body to power these tools. But we now know that the human tool maker and user shares some of these abilities with many other creatures.

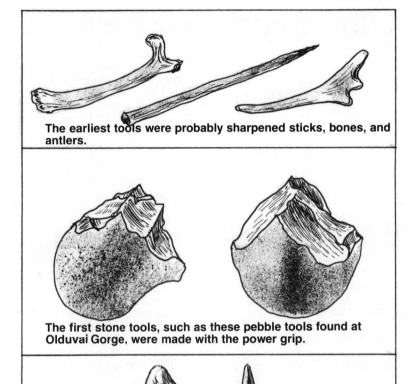

The earliest tools were probably sharpened sticks, bones, and antlers.

The first stone tools, such as these pebble tools found at Olduvai Gorge, were made with the power grip.

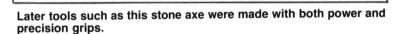

Later tools such as this stone axe were made with both power and precision grips.

Glossary

Ancestor — A person or animal from whom another comes down.

Bower — A decorated nest.

Computer — A machine that uses information given to it to solve problems.

Crop — A storage place for food inside the bodies of some animals.

Crustacean — An invertebrate that breathes with gills and has a hard shell. Most live in the ocean.

Entomologist — A scientist who studies insects.

Ethologist — A scientist who studies animals in the wild.

Ethology — The study of animals in the wild.

Generation — The average length of time between the birth of parents and the birth of their children. A generation is usually 20 to 25 years.

Gill — A part of the body of some animals that enables them to breathe in water.

Grooming — An activity in which primates clean the hair of other primates.

Imitate — To copy.

Insight — A kind of learning in which a person or animal suddenly sees the connection between two things.

Invertebrate — An animal without a backbone.

Larvae — A life stage of some insects. Many larvae look like worms.

Mammal — Vertebrate animals that have hair and feed their young with milk.

Primate — A special group of mammals made up of humans, apes, and monkeys.

Psychologist — A scientist who studies the brain and behavior of people and animals.

Species — A group of closely-related animals. Members of a species can mate with one another and produce young that can also mate and produce young.

Tool — An object outside the body that helps an animal use its legs or some other body part to achieve a goal. A tool is usually picked up and carried around before being used.

Vertebrate — An animal with a backbone.

Index

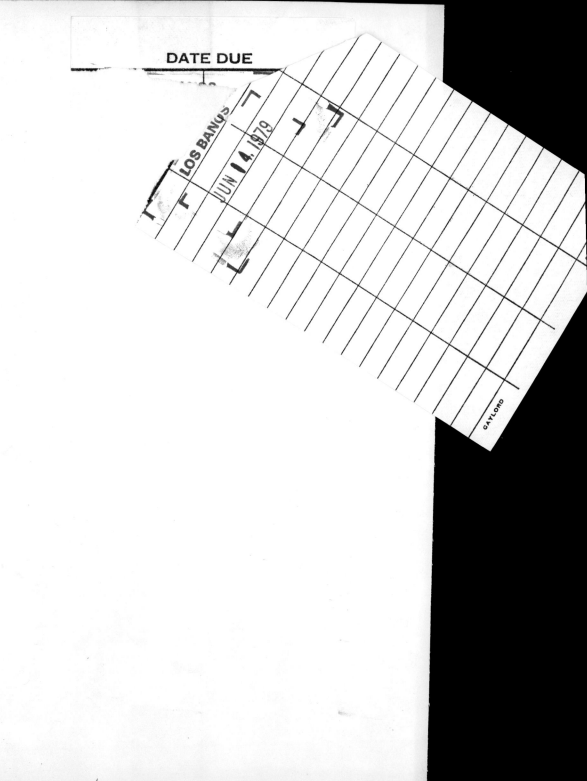

GAYLORD